YOUR KNOWLEDGE HAS VALUE

Rehan Jamil

Development Tendency of Energy. A Short Review

GRIN Verlag

Bibliografische Information der Deutschen Nationalbibliothek:

Die Deutsche Bibliothek verzeichnet diese Publikation in der Deutschen National-
bibliografie; detaillierte bibliografische Daten sind im Internet über http://dnb.d-
nb.de/ abrufbar.

Dieses Werk sowie alle darin enthaltenen einzelnen Beiträge und Abbildungen
sind urheberrechtlich geschützt. Jede Verwertung, die nicht ausdrücklich vom
Urheberrechtsschutz zugelassen ist, bedarf der vorherigen Zustimmung des Verla-
ges. Das gilt insbesondere für Vervielfältigungen, Bearbeitungen, Übersetzungen,
Mikroverfilmungen, Auswertungen durch Datenbanken und für die Einspeicherung
und Verarbeitung in elektronische Systeme. Alle Rechte, auch die des auszugsweisen
Nachdrucks, der fotomechanischen Wiedergabe (einschließlich Mikrokopie) sowie
der Auswertung durch Datenbanken oder ähnliche Einrichtungen, vorbehalten.

Imprint:

Copyright © 2013 GRIN Verlag GmbH
Druck und Bindung: Books on Demand GmbH, Norderstedt Germany
ISBN: 978-3-656-59434-5

GRIN - Your knowledge has value

Der GRIN Verlag publiziert seit 1998 wissenschaftliche Arbeiten von Studenten, Hochschullehrern und anderen Akademikern als eBook und gedrucktes Buch. Die Verlagswebsite www.grin.com ist die ideale Plattform zur Veröffentlichung von Hausarbeiten, Abschlussarbeiten, wissenschaftlichen Aufsätzen, Dissertationen und Fachbüchern.

Visit us on the internet:

http://www.grin.com/

http://www.facebook.com/grincom

http://www.twitter.com/grin_com

Development Tendency of Energy: A Short Review

Rehan Jamil, Irfan Jamil, Ming Li and Zhao Jinquan

Abstract—Energy is the important source for the development of the society and it's the basic support of national economy and the base for human living. As the development of economy, abrupt increase of population and continuous improvement of living standards, the demand of energy increases continuously, which caused the impetuous scramble of energy source in the world, and urged the attention of the countries for current status and development trends of energy.

Keywords—Energy, Energy Supply Situation, Energy Production & Consumption.

I. INTRODUCTION

IN 2007, the consumption of primary energy increased by 2.4%, although it's a little lower than the increase of 2.7% in 2006, it's still the 5th successive years higher than the average level [3]. The consumption increase of energy in the Asia-Pacific region is 2/3 of the increase of the world, and it's increasing continuously with 5% higher than the average level. However, the consumption of energy decreased by 0.9% in Japan. In North America, the consumption recovered in 2006 and rebound with an increase of 1.6% which was double of the average for last 10 years. In China, the increase rate of energy consumption was 7.7% in 2006, although it's higher than the average of last ten years (same situation of economy increase in the same period), it's the lowest increase rate since 2002. The consumption of energy once again took half of the increase of the world. The increase rate of energy consumption in India was 6.8%, ranking the third after China and England. In European Union the consumption of energy decreased by 2.2%, in which Germany was the country with maximum decrease rate in the world [1]-[3].

II. ANALYSIS OF ENERGY SUPPLY SITUATION

A. Petroleum

In 2007, the consumption of petrol increased by 1.1% in the world, a little lower than the average level of the last 10 years.

The consumption of petrol in the petrol export region, i.e. Middle East, South America, Central America and Africa are 2/3 of the increase of petrol consumption of the world. In Asia-Pacific Region, the increase rate was 2.3%, which is almost at the same level as historical average. Although the increase of petrol consumption in China and Japan is lower than the average level, there is much increase in other emerging economies. The consumption of the countries of the Organization for Economic Co-operation and Development decreased by 0.9%, the great demand made the increase of naphtha distillate oil the same as mesoplasm distillate oil for the first time since 2002 [3], [9].

In 2007, yield of petrol decreased by 0.2% in the world with 130,000 barrels/day. It's the decrease for the first time since 2002. The production of OPEC decreased by 350,000 barrels each day in 2007 due to the impact of production reduction in November of 2006 and February of 2007 [3].

Although the production was reduced by OPEC and the consumption increased in petrol export countries, the trade of international crude oil and processed oil increased. The increase mainly came from processed oil, it can be seen that the oil refining system is under unbalance and limited status.

B. Natural Gas

The reserves amount of natural gas in the world is 208.4 trillion m^3, which will meet the demands of the world for 63.6 years.

In 2007, the consumption of natural gas increased by 3.1% in the world, more than the average increase in the past. The main reason is that the increase in North America, Asia-Pacific Region and Africa is more than the average of the world. The demand for deep winter and power generation with natural gas caused that the increase of natural gas consumption accounted for almost half of the increase in the world. At the same time, natural gas almost led all the increase of energy consumption in America. The increase rate of natural gas was 19.9% in China, ranking the second in the world. Because of warming winter, the consumption of EU decreased by 1.6%, decrease for two successive years [4].

In 2007, the production of natural gas increased by 2.4%, the same as the situation of consumption. America is the country with maximum increase of natural gas supply and increase rate is 4.3%, the maximum increase since 1984. The production of natural gas decreased by 6.4% in European Union, in which England was the top one for two successive years-9.5% decrease. The production increased much in Former Soviet Union region except Russia which balances out the little decrease in Russia. The production of natural gas increased by 18.4% and 17.9% in China and Qatar respectively which rank the second and the third.

In 2007, international natural gas trade was continuously inactive with an increase rate of 2.3% which was lower than half of the average level of last ten years. As the low consumption in Europe, the exported natural gas through

Rehan Jamil is with the School of Physics & Electronic Information, Yunnan Normal University, Kunming, China (phone: +86-18388144878; e-mail: ch.rehan.jamil@gmail.com).

Irfan Jamil is with the College of Energy & Electrical Engineering, Hohai University, Nanjing, China (e-mail: irfan.edu.cn@gmail.com).

Ming Li is with the School of Energy & Environmental Science, Yunnan Normal University, Kunming, China (e-mail: lmllldy@126.com).

Zhao Jinquan is with the College of Energy & Electrical Engineering, Hohai University, Nanjing, China (e-mail: zhaojinquan@hhu.edu.cn).

piping was suspended. The liquefied natural gas (LNG) transportation increased by 7.3% due to the lasting export increase in Qatar and Nigeria, the same as the historical increase level. The integration of LNG trade was lasting in Atlantic Ocean and Pacific Ocean area. As the price for actual goods higher than the one in Europe, the freighter turned to America, LNG receipt increased by 1/3 in America.

C. Coal

The reserves amount of coal in the world is 860.938 billion tons, which will meet the demands of the world for 112 years.

Coal increased fastest for five successive years. Coal consumption increased by 4.5% in the world which is higher than the average level 3.2% of last ten years. And the increase distributed widely, except Middle East, the other region had the increase more than the average of last ten years. Coal consumption increased by 7.9% in China which is the lowest since 2002, but it's still more than 2/3 of the increase in the world. The consumption increased by 6.6% in India, the increase rate in the countries of economy cooperation and development organization is 1.3%, all are higher than the average of last ten years [4], [5].

D. Nuclear Power

Until the end of 2007, it is confirmed the exportable uranium mine reserves is 5.4688 million tons with price of 130 dollars per kilogram; the uranium mine production in 2007 is about 41.165 thousand tons. It is anticipated the R/P of uranium mine is equivalent to 100 years, while the service life will increase 50 times if the nuclear fuel is reclaimed and reused after Special treatment [4].

Nuclear power decreased by 2% which made the maximum decrease record. But more than 90% are from Germany and Japan-Because of the shutdown of largest nuclear power plant in the world due to earthquake. Hydropower increased by 1.7%, a little bit higher than the average of last ten years. The increase of installed capacity in China and Brazil as well as the increase of hydropower availability in Canada and North Europe balance out part of the decrease of hydropower in America and South Europe due to drought.

E. Renewable Energy

Renewable energy is only a small part of energy consumption in the world, but in 2007, there was a rapid increase of renewable energy. Such as the production of ethanol increased by 27.8%. The installed capacity of wind energy and solar energy increased by 28.5% and 37% respectively, keeping with the same level in the history [5].

As for hydraulic power, according to the international energy resources investigation in 2001, theoretically the annual hydraulic capacity in the world is 40 trillion 704 billion kilowatt-hour; however the technical utilized capacity in practice is about 14 trillion 379 billion kilowatt-hour. However the gross power generation amount of the whole world in 2006 is about 777 Billion watts and provides 2.998 trillion kilowatt-hours electricity [5], [6].

III. STATUS QUO AND CHARACTERISTICS OF GLOBAL ENERGY CONSUMPTION

A. Influenced of Economy Development and Population Increase, Primary Energy Consumption Amount up Continuously

As the global economy dimension expands continuously, global energy consumption constantly increases. GDP of the whole world in 1990 was 26.5 trillion dollars (calculated supposing the price index did not change in 1995); in 2000 the number reached 34.3 trillion, with annual increase 2.7%. According to BP Energy Statistics in 2011, the global primary energy consumption in 1973 was only 5.73 billion tons oil equivalent, while the amount reached 12.274 billion tons oil equivalent in 2011 [2],[6]. The annual growth rate of global energy consumption in the past 30 years is about 1.8%. Fig. 1 and Fig. 2 shows growth & distribution of global primary energy consumption.

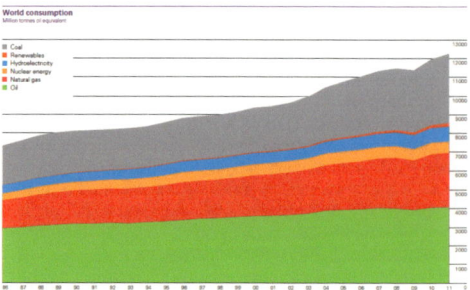

Fig. 1 Growth of global primary energy consumption source: BP Energy Statistics 2012 (Unit: Million Tons Oil Equivalent)

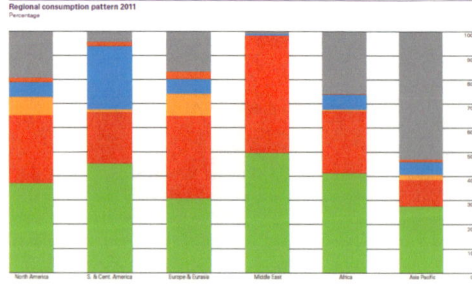

Fig. 2 Distribution of global primary energy consumption

B. Global Energy Consumption Presents Different Growth Modes; the Growth Rate of Developed Countries Is Obviously Lower than That of Developing Countries

In past 30 years, the gross energy consumption amount increases continuously, however in North America and Europe where economy, technology and society is comparatively developed, the growth rate of energy consumption is quite slow. The energy consumption percentage of these regions compared with the whole world also decreases annually. North America decreases from 35.1% in 1973 to 22.6% in 2010, while Europe decreases from 42.8% in 1973 to 23.8% in 2010 [8].

C. Main reason

1) Economy development of developed countries has entered post-industrialization phase, whose industrial structure turns into low-consumption but high production. Highly consuming manufacturing industry is transferred to developing countries.
2) Developed countries emphasize energy conservation and raising energy utilization efficiency.
3) Global energy consumption structure tends to optimize, while regional difference is obvious.

D. Global Energy Consumption Structure Tends to Optimize, While Regional Difference Is Obvious

Since the industrial revolution in 1870's, consumption of fossil fuel increases sharply. At the preliminary stage, coal is the main fuel. Fig. 3 shows global distribution of petrol production and consumption. After 1990, especially since WWII, production and consumption of petrol and natural gas rises ceaselessly. Fig. 4 shows global distributions of natural gas production and consumption. Petrol surpasses coal and takes the lead of primary energies in 1960's.

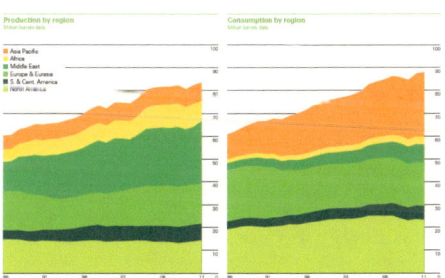

Fig. 3 Global distribution of petrol production and consumption

Although the world goes through two petroleum crises in 1970's, global petrol consumption presents no tendency of decreasing. After that, the percentage petrol and coal takes decreases slowly and percentage of natural gas rises. Fig.5 show global distribution of coal production and consumption. At the same time, other forms of new energies such as nuclear power, wind energy, hydro-energy and geothermal energy is gradually developed and utilized, thus a new energy structure comes into being, in which fossil fuels take the lead and reproducible energies, new energies coexist. By the end of 2011, fossil energies are still the main energies in the world, with the percentage 87.1% among all the primary energies [5], [8]. While within fossil energies, the percentage of petrol is 33.1%, coal 30.3%, natural gas 23.7%. Although Non-fossil energies and reproducible energies increase fast, their

percentage is relatively small, about 12.9%. Figs. 6 and Fig. 7 show global distribution of nuclear power and hydro energy production and consumption.

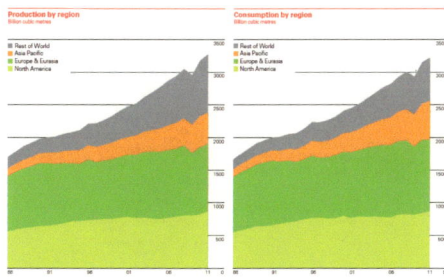

Fig. 4 Global distribution of natural gas production and consumption

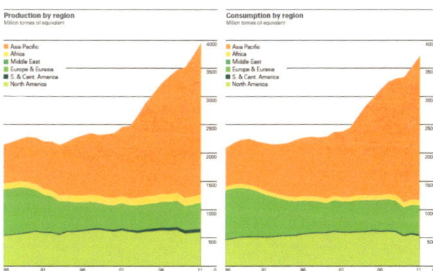

Fig. 5 Global distribution of coal production and consumption

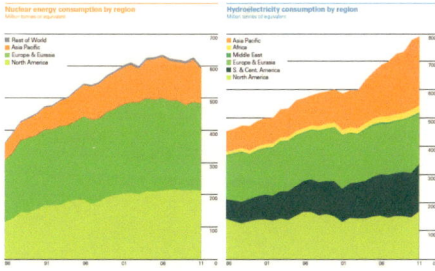

Fig. 6 Global distribution of nuclear power & hydro energy production and consumption

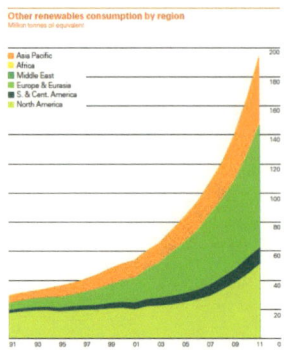

Fig. 7 Global distribution of nuclear power & hydro energy production and consumption

E. Main Reason Global Energy Consumption Structure Tends to Optimize, While Regional Difference Is Obvious

As energy resources of some regions in the world dry up, energy transaction between different countries and regions will further increase, which causes great demand of energy transportation as well. Issues such as energy storage and transportation or energy supply security are emphasized increasingly. Figs. 8 and 9 show global energy trade in 2011.

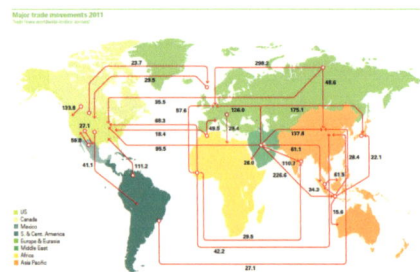

Fig. 8 Global energy trade in 2011

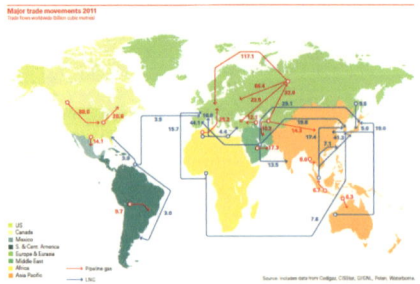

Fig. 9 Global energy trade in 2011

IV. ANALYSIS OF THE TREND OF ENERGY SUPPLY AND DEMAND IN THE WORLD

According to the latest forecast of Energy Information Administration of America (EIA), as the developing of economy and society in the world, the demand of energy will increase continuously in the world. It's forecasted that the demand of energy will reach 12.89 billion tons in 2020, and 13.650 billion tons in 2025, the average annual increase rate will reach 1.2%. In the two developed region, Europe and North America, the proportion of energy consumption will decrease, but in Asia, Middle East, Middle and South America regions, it will keep increasing. As the increase of energy storage concentration in the world, the scramble of energy source will be more and more serious, and the measures will be more complicated. It's still possible to arise of dispute and war caused by scramble of energy. Mr. Fan Dewei, CEO of Shell Group, once pointed out three austere facts that make the energy system in turbulence in the world in recent decades. The first fact is that the increase of energy demand is accelerating. As the population will increase to 9 billion from 6 billion in next 40 years in the world, the energy consumption may reach double of the one at present or even more in 2050 [4], [7], [8].

The second serious fact is that the normal petrol and natural gas easy to be exploited can't satisfy the rapid increase of demand. It is estimated by International Energy Agency (IEA) that there are 2,000 billion barrels of oil equivalent petroleum and natural gas reserves, including normal energy and abnormal energy, such as oil shale and oil sand. Theoretically, according to the speed of present consumption, this energy source is enough for human being for about 400 years. But actually, less than half of the energy can be exploited with present technologies.

The third serious fact is that more coal consumption may cause more CO_2 emission than the acceptable limit. IEA forecasts that coal consumption may increase by 60% in the next 20 years. All the countries favor in consideration of the energy source safe, reducing reliance on import of petroleum and natural gas. But CO_2 emission from coal-fired power generation is double than the one from nature gas-fired power generation. According to the statistics of EIA, CO_2 emission was about 21.56 billion tons in 1990, it reached 23.9billion tons in 2001, and 27.72 billion tons in 2010, it is forecasted to reach 37.12billion tons in 2025, the average annual increase rate is 1.85%. Facing to these serious challenges, the energy supply and demand in the future has to develop toward the direction of diversification, clean, high efficient, globalization and marketization [10].

A. Diversification

After firewood-dominant, coal-dominant and petroleum-dominant times, current global energy consumption structure with natural gas as the main re-source is under formation. Meanwhile, water energy, nuclear energy, wind energy and solar energy are broadly utilized. A diversified development structure of global energy is established as a result of sustainable development, environmental protection, the cost of energy supply and the change of available energy structure. Natural gas consumption will be stably increased. In some regions, coal-fired power plants are inclined to be replaced by gas-fired power plants. In future, while developing conventional energy, people will pay more attention to the development of new energy and renewable energy. In the Renewable Energy Development Plan of 2010 of the European Union, 40,000,000kW wind power and 105,000,000kW hydropower has developed in 2010 [8], [9]. A new energy strategy was established in the <<Energy White Paper>> issued by the British government in early 2003, i.e. compared with the current proportion of 3%, the output of renewable energy generated power shall account for 15% of the total power output of Britain in 2015, and 20% in 2020.

B. Cleanness

As new technology of global energy upgrades and environmental protection standard becomes strict, global energies in the future will develop towards the direction of cleanness. Energy productive process is required clean. And energy industry is also required to produce more qualified clean energies, thus the percentage clean energies in gross energy consumption will increasingly grow. With the progress of global new energy technology development and stricter environmental protection standards, clean energy will be mainly developed in the world in future, which requires not only clean energy production process, but also the continuous production of more and better.

Clean energy by energy industries. Clean energy consumption will account for a higher percent in the total energy consumption. It is expected that in 2025, the global energy consumption proportioning will have the following change, i.e. coal will be reduced to 21.72% from the current 26.47%, natural gas will be increased to 28.40% from the current 23.94%, and petroleum will be maintained at 37.60%~37.90%. Meanwhile, clean utilization of coal which was deemed to be dirty energy in the past and traditional energies such as firewood, stalk and excrement will be developed, and with breakthrough on the R&D, clean coal technology (such as coal liquefaction technology, coal gasification technology, coal desulphurization and deduct technology), biogas technology and bio-diesel technology will be broadly utilized. Some countries, such as France, Austria, Belgium and Netherlands have closed all their domestic coal mines, so as to develop nuclear power [9].

C. High-efficiency

There is a great potential for the improvement of energy utilization efficiency due to the big difference between energy processing efficiency and consumption efficiency. With the progress of global new energy technology development, the global energy utilization efficiency in future will be gradually improved, and the energy intensity will be gradually reduced. For example, when being counted with the constant price of USD in 1997, the global energy intensity in 1990 was 0.3541 ton oil equivalent/thousand USD, and reduced to 0.312 ton oil

equivalent/thousand USD in 2001. It was expected that, it will be reduced to 0.2759 ton oil equivalent/thousand USD in 2010 and 0.2375 ton oil equivalent/thousand USD in 2025 [8].

However, energy intensities of various regions in the world differ a lot. For example, the energy intensity of developed countries in the world was only 0.2109 ton oil equivalent/thousand USD in 2001, while the energy intensity of developing countries in 2001~2025 was expected to be 2.3~3.2 times of that of the developed countries, from which we can see there is a great potential for global energy conservation.

D. Globalization

Due to the non-uniformity of global energy resource distribution and demand distribution, it is getting increasingly difficult for countries and regions to meet their energy demand with domestic resources only. They need resources supplied by other countries and regions, resulting in the trend of greater global trading volume and increasing trade quotas. Taking petroleum trade as an example, the global petroleum trading volume was increased to 2,120,000,000t in 2000 and 2,180,000,000t in 2002 from 1,220,000,000t in 1985, with the annual average growth rate (AAGR) at about 3.46%, which is higher than the AAGR of global petroleum consumption of the corresponding period, i.e.1.82%. In the foreseeable future, the global net petroleum import quantum will be gradually increased, with the AAGR of 2.96%. It is expected that it will reach 29,300,000 barrels/day in 2010, 40,800,000 barrels/day in 2020, and 48,500,000 barrels/day in 2025. The globalization of global energy supply and consumption will be accelerated, and the main energy production countries and energy consumption countries in the world will actively participate in the globalization progress of energy demand and supply market.

E. Marketization

Marketization is the optimum mode for international energy resource optimizing configuration and utilization. With the development of global economy, especially the speedup of marketization reformation progress of countries, the marketization degree of global energy utilization becomes higher and higher. Governments in the world directly interfering with energy utilization will be less and provide more services for the energy market. Particularly, governments will complete energy laws and regulations of their countries and regions, and provide good energy market environment. Currently, countries with rich energy resources such as Russia, China, Kazakhstan and Libya are completing their national energy investment policies and administrative management measures. In this way, the marketization degree and normalization degree of energy production of these countries will be increased and bringing advantages for overseas investors.

V. CONCLUSION

As a result of the development of global economy, especially the fast growing of the economies in China and

India, global energy becomes in short supply. Meanwhile, human beings have realized that energy utilization will cause severe environmental impact. More types of energy and technologies are needed to meet the future demand and reduce environmental damage. Therefore, the most of new energy technologies trend to be used to develop and utilize clean and renewable energy.

ACKNOWLEDGMENT

This work is funded by grants from the National and International Scientific and Technological Cooperation Projects of China (Grant number: 2011DFA60460) and material support from China Power Investment Yuanda Environmental Protection Engineering Co., Ltd.

REFERENCES

[1] Zhang Z X. Why did the energy intensity fall in China's industrial sector in the 1990s, Energy Economics, 2003, 25(6): pp. 625-638.
[2] IEA, "World Energy Outlook 2011", Paris: International Energy Agency, 2011.
[3] Wu Yong, Liu Changbin, Study on motivation policies for building energy-conservation economy, Beijing, China Construction Industry Press,2007
[4] Yin, J.-H., Wang, Z.-H.: The Relationship between Energy Consumption and Economic Growth in China–Based on the Data in the Period of 1953-2008. J. Science Research Management vol. 7, pp 122–129. 2011.
[5] Pehnt, "Marin. Dynamic lifecycle assessment (LCA) of renewable energy technologies", Renewable Energy, vol. 31, no. 1, pp. 55-71, January 2006.
[6] Research Center of Economic Security of the Chinese Academy of Contemporary International Relations, A big picture of Global Energy, Press, 2005.
[7] Schainker, R. B. "Executive Overview: Energy Storage Options for a Sustainable Energy Future". IEEE Power Engineering Society General Meeting, Vol. 2, pp: 2309-2314.2004.
[8] IEA, "Energy Technology Perspectives 2010: Scenarios & Strategies to 2050", Paris: International Energy Agency, 2010.
[9] RFA, "2010 world fuel ethanol production", Washington, USA: Renewable Fuels Association, 2011.
[10] B. K. Sovacool, "Valuing the greenhouse gas emissions from nuclear power: A critical survey", Energy Policy, vol. 36, no. 8, pp. 2950-2963, August 2008.

Rehan Jamil was born in Punjab province, City Multan, Pakistan on Feb 25, 1987. He received his bachelor in B.Sc. Electrical (Electronic) Engineering from Federal Urdu University of Arts, Science & Technology Islamabad Pakistan in 2009. Currently he is pursuing his Master degree at Yunnan Normal University, Kunming China. His research interest involves in Electronics, Renewable energy power generation. He is member of National Society of Professional Engineers (NSPE) and also member of American Society of Mechanical Engineer (ASME).

Irfan Jamil was also born in Punjab province, City Multan, Pakistan on Feb 25, 1987. He received his bachelor degree in Electrical Engineering and its Automation from Harbin Engineering University, Harbin, China in 2011. Currently he is pursuing his Master degree at Hohai University, Nanjing, China. During these days he is doing master research as a Visiting Research Scholar at Tsinghua University, Beijing China. His research interest involves in Power electronics and Power system Automation. He is member of National Society of Professional Engineers (NSPE) and also member of American Society of Mechanical Engineer (ASME). He is also member of Chinese Society for Electrical Engineer (CSEE).

Prof. Dr. Ming Li was born in Yunnan Province, Kunming, China on 20 October 1964. He received his B.E and M.E degrees in Power Engineering from Shanghai University of Science & Technology, Shanghai China, in 1985 and 1991, in 2000 obtained his Ph.D. degree in Power & Energy Engineering from Shanghai Jiao tong University, Shanghai, China. From 2001 to 2002, he was as a postdoctoral associate in College of Science, Solar energy Research Institute of Shanghai Jiao tong University. From 2003 to 2004, in Australia National University renewable energy Research Center as a visiting scholar. He was a Professor in School of Physics & Electronic information at Yunnan Normal University, Kunming, from 2000 to 2007. Currently he is Dean, School of Energy & Environmental Science and also Director, Solar Energy Research Institute of Yunnan Normal University, Kunming, China. He has been Published more than 60 research papers in International Journals and Conference papers. He is Editor of the Chinese Journal of Solar Energy & Vice President of Yunnan Solar Energy Association. His research interests in the area of solar thermal energy utilization, solar refrigeration, solar building, solar concentrating heating & Power system, refrigeration & cryogenics, heat & mass transfer.

Prof. Dr. Zhao Jinquan was born in Yangquan, Shanxi province, China, on June 26 1972. He received B.S. and Ph.D. degrees, all in electrical engineering, from Shanghai Jiao tong University, Shanghai, China, in 1993 and 2000, respectively. From 1993 to 1995, he was an engineer in Guangzhou Power Company, Guangzhou, China. From December 2000 to September 2003, he was a postdoctoral associate in Cornell University, Ithaca, NY. He was a postdoctoral associate in Tsinghua University, Beijing, China. Currently he is a professor in Electrical Engineering department, Hohai University, Nanjing, China. He has been published more than 28 papers in many international conferences. His research interests in the area of voltage stability analysis and control, OPF and its applications.